Skander Kacem

Evaluation of Interference Cancellation Architectures for Heterogeneous Cellular Networks

Anchor Compact

Kacem, Skander: Evaluation of Interference Cancellation Architectures for Heterogeneous Cellular Networks. Hamburg, Anchor Academic Publishing 2015
Original title of the thesis: «buchtitel»

Buch-ISBN: 978-3-95489-343-0
PDF-eBook-ISBN: 978-3-95489-843-5
Druck/Herstellung: Anchor Academic Publishing, Hamburg, 2015

Bibliografische Information der Deutschen Nationalbibliothek:
Die Deutsche Nationalbibliothek verzeichnet diese Publikation in der Deutschen Nationalbibliografie; detaillierte bibliografische Daten sind im Internet über http://dnb.d-nb.de abrufbar

Bibliographical Information of the German National Library:
The German National Library lists this publication in the German National Bibliography. Detailed bibliographic data can be found at: http://dnb.d-nb.de

© Anchor Academic Publishing, ein Imprint der Diplomica® Verlag GmbH
http://www.diplom.de, Hamburg 2015
Printed in Germany

Abstract

With the increasing data throughput requirements, the cellular network needs to move from homogeneous to heterogeneous system. In fact, the coexistence of different types of base stations with different capabilities such as femto/pico base stations as well as relays and macro base stations in random placements should improve the coverage and the spectral efficiency of the cellular networks.

However, the complexity of inter-cell interference management will grow drastically and traditional interference avoidance/mitigation approaches need to be revised.
Approaching this problem at the user equipment (UE), is of great interest since it can rely on little coordination among base stations.

The work presented in this thesis focuses on a downlink interference cancellation at the UE and shows that such an intelligent receiver can bring its promised benefit only if the base stations get involved in the interference cancellation, specifically in the channel estimation process. The limitations of this approach are evaluated and depending on the surrounding base stations two solutions are proposed and discussed.

Zusammenfassung

Aufgrund steigender Anforderungen an Datendurchsatzkapazität muss sich das Mobilfunknetz von einem homo- zu einem hetero-genen System entwickeln.

Das Zusammenspiel verschiedener Arten von Basisstationen mit unterschiedlichen Fähigkeiten bzw. Sendeleistungen, wie etwa Femto/Pico- oder Relais und Macro-Basisstationen mit zufälliger Verteilung, sollte die Netzabdeckung und spektrale Effizienz des netzes verbessern.

Allerdings, die Komplexität von Inter-Cell-Interference-Management wird stark ansteigen, weshalb die traditionellen Interferenz-Vermeidungs und -verringerungsansätze überarbeitet werden müssen.

Dieses Problem auf Ebene des Endgerätes (UE) anzugehen, würde mit wenig Koordination zwischen den Mobilfunkzellen auskommen.

Diese Arbeit konzentriert sich auf eine Downlink-Interference-Cancellation auf UE Ebene und zeigt, dass ein solch intelligenter Empfänger die erwarteten Vorteile nur dann erreichen kann, wenn die Basisstationen in die Interference-Cancellation, insbesondere by der Kanalschätzung am UE, einbezogen werden.

Die Grenzen dieses Ansatzes werden untersucht und abhängig von den umgebenden Basisstationen werden zwei Lösungen vorgeschlagen und evaluiert.

Contents

Notations   **9**

Acronyms   **10**

1 Introduction   **12**
  1.1 Motivation and Objective 12
  1.2 Outline and Organization of the Thesis 12

2 Fundamentals   **14**
  2.1 Bit-Interleaved Coded Modulation 14
     2.1.1 System and Signal Model 14
     2.1.2 Low Complexity LLR Metrics for BICM Receivers 16
  2.2 Interference-Aware System 19
     2.2.1 IA Receiver: System & Signal Model 19

3 Structure of the Simulator   **22**
  3.1 System Parameters & General Code Structure 22
  3.2 Radio Propagation Channel 25
     3.2.1 Simulation of AWGN-Channel model 25
     3.2.2 Path Loss Channel Model 28
     3.2.3 Simulation of Rayleigh Fading Channel 29
  3.3 The Baseband Part of the Transmitter 32
     3.3.1 Convolutional Coding and Puncturing 33
     3.3.2 Bit-Interleaver 34
     3.3.3 Bit-Level Scrambling 35
  3.4 Interference Model 36
  3.5 The Baseband Part of the Receiver 38
     3.5.1 Pilot-based Channel Estimation 38
     3.5.2 De-Puncturing and Soft Output Viterbi Decoding 41
     3.5.3 Metric Computing Device 42
  3.6 Base Stations Channel Estimation Enhancement 46
     3.6.1 Serving Base Station: Holes 48

Contents

        3.6.2   Interfering BS: Pilot Boosting   48

4   **Simulation Results**   **51**

5   **Summary and Outlook**   **57**

Bibliography   **59**

List of Figures

2.1 Block diagram of BICM transmission: encoder ENC, bit interleaver π, modulator $\mathcal{M}$, metric computing device $\mathcal{M}^{-1}$, deinterleaver π^{-1} and decoder DEC 14

2.2 Binary labelling sets for QPSK and 16-QAM 15

2.3 Co-channel interference between hexagonal cells with a frequency reuse of factor one 19

3.1 Structure of the Simulation Testbed 23

3.2 Illustration of one LTE Slot 24

3.3 Power Spectrum of LTE DL Signal 24

3.4 AWGN Channel model 25

3.5 generated AWGN, where $z \in \mathcal{CN}(0, 0.35)$ 26

3.6 AWGN-Channel model: Impact on one OFDM symbol 27

3.7 BER theoretical vs simulated 28

3.8 Rayleigh Channel Model 30

3.9 Non Line of Sight (NLOS)- Multipath 30

3.10 Rayleigh Distribution and Example 31

3.11 BER performance in a Rayleigh Propagation Channel 31

3.12 Block Diagram of the Baseband Transmitter 32

3.13 OFDM Modulation, OFDM Symbol 32

3.14 Encoder with generator polynomials $g_1 = [111]$ and $g_2 = [101]$ 33

3.15 Trellis diagram of $(7, 5)_8$ encoder 34

3.16 Interference Model 36

3.17 Baseband Part of the Receiver 38

3.18 Pilots within a Subframe 39

3.19 Channel Estimate of a 16-taps Rayleigh Channel by SNR = 15 dB 40

3.20 $2 \times$ 1-D Interpolation/Extrapolation 41

3.21 Linear Interpolation in the Frequency-Domain 42

3.22 Metric Computing Device: Metric as Log Likelihood Ratio 42

3.23 LLR vs Bit 43

3.24 1tap Rayleigh Channel effects on an OFDM Symbol by $SNR = 5$ dB . . . 44

3.25 OFDM symbol after usual Phase and Amplitude Equalization 44

3.26 OFDM symbol after Matched Filter 45

3.27 $\hat{H}_1$ vs. SIR 47
3.28 $\hat{H}_2$ vs. SIR 47
3.29 IAR Performance Boundaries 48
3.30 Serving BS enhancement: Holes at the pilot positions of the interfering
     signal 49
3.31 Interfering BS: Pilot Boosting 50

4.1 BER performance of IA-R vs. II-R by SNR = 5 dB and variable SIR and
     Perfect Channel Knowledge 51
4.2 BER performance of IA-R vs. II-R by SNR = 20 dB, variable SIR and
     Perfect Channel Knowledge 52
4.3 IA-R vs. II-R by Perfect Channel Knowledge 53
4.4 BER performance of IA-R vs. II-R by SNR = 15 dB and variable SIR
     and LS Channel Estimation 54
4.5 IA-R vs. II-R by LS Channel Estimation 54
4.6 IA-R Performance with Base Station Interference Cancellation Enhance-
     ment, by SNR = 5 dB and variable SIR 55
4.7 IA-R Performance with Base Station Interference Cancellation Enhance-
     ment, by SNR = 20 dB and variable SIR 55
4.8 Proposed Interference Cancellation Architectures performance by variable
     SNR/SIR values 56

Notations

arg	operator that delivers the argument
$*$	convolutional operator
$\circledast_N$	circular convolution with block length N
$(\hat{\cdot})$	the estimate of a parameter or a parameter vector/-matrix
$E\{\cdot\}$	expectation
$\Im\{\cdot\}$	imag. part of argument
$\log_n$	logarithm to base n
$\mathbf{A}$	bold upper case indicates a matrix.
max	maximum of arguments
min	minimum of arguments
$(\cdot)^*$	complex conjugate
$Pr\{\cdot\}$	probability
$(\cdot)^\dagger$	pseudo-inverse
$\Re\{\cdot\}$	real part of argument
$\mathbf{a}$	bold lower case indicates a vector.
$(\cdot)^T$	transpose

Acronyms

3GPP	third generation partnership project
ARQ	automic repeat request
AWGN	additive white Gaussian noise
BER	bit error rate
BICM	bit-interleaved coded modulation
BPSK	binary phase shift keying
BS	base station
BW	bandwidth
CIR	channel impulse response
CP	cyclic prefix
CRC	cyclic redundancy check
DFT	discrete Fourier transform
DL	downlink
eNodeB	evolved Node B
FEC	forward error correction
i.i.d	independent and identically distributed
IA-R	interference-aware receiver
ICI	inter-channel interference
IDFT	inverse discrete Fourier transform
II-R	interference-ignorant receiver
ITU-R	international telecommunication union, radiocommunication sector
LLR	log likelihood ratio
LTE	long term evolution

MAP	maximum a posteriori probability
MF	matched filter
ML	maximum likelihood
MS	mobile station
MSE	mean squared error
NLOS	non line of sight
OFDM	orthogonal frequency division multiplex
pdf	probability density function
PHY	physical layer
PL	path loss
QAM	quadrature amplitude modulation
QPSK	quadrature phase-shift keying
RB	resource block
RE	resource element
RS	reference signal
Rx	receiver
SINR	signal-to-interference and noise ratio
SIR	signal-to-interference ratio
SISO	single-input single-output
SNR	signal-to-noise ratio
TB	transport block
TTI	transmission time interval
Tx	transmitter
UE	user equipment
XOR	exclusive or

1 Introduction

1.1 Motivation and Objective

With the increasing data throughput requirements, the cellular network needs to move from homogeneous to heterogeneous system. In fact, the coexistence of different types of base stations with different capabilities such as femto/pico base stations as well as relays and macro base stations in random placements should improve the coverage and the spectral efficiency of the cellular networks.

However, the complexity of inter-cell interference management will grow drastically and traditional interference avoidance/mitigation approaches need to be revised.
Approaching this problem at the user equipment (UE), is of great interest since it can rely on little coordination among base stations.

The work presented in this thesis focuses on a downlink interference cancellation at the UE and shows that such an intelligent receiver can bring its promised benefit only if the base stations get involved in the interference cancellation, specifically in the channel estimation process. The limitations of this approach are evaluated and depending on the surrounding base stations two solutions are proposed and discussed.

1.2 Outline and Organization of the Thesis

This thesis consists of four chapters, that are organized as follows:
Chapter 2 first introduces the basic concepts for the understanding of the thesis. It defines the bit-interleaved coded modulation (BICM), the low complexity log-likelihood (LLR) metrics for BICM receiver and the interference-aware receiver (IA-R).

In **Chapter 3**, the structure of the simulator is described. This includes transceiver system model as well as the propagation channel and the interference model. Furthermore, two solutions for IA-R performance enhancement are proposed.

Chapter 4 contains the simulation results of the considered systems. First, a comparison between the IA-R and the interference-ignorant receiver II-R under perfect channel knowledge are evaluated, providing an estimate of the upper performance bound of the IA-R. Then, the performance of the IA-R under more realistic assumptions, with and without base station interference cancellation enhancement, is tested.

Finally, **Chapter 5** summarizes the thesis and gives a brief outlook on future research and developments.

2 Fundamentals

2.1 Bit-Interleaved Coded Modulation

2.1.1 System and Signal Model

Bit-interleaved coded modulation (BICM) is the serial concatenation of a channel encoder, a bit-wise interleaver π and an M-ary modulator, which maps blocks of m bits to a QAM constellation x from a symbol set $\mathcal{X} \subseteq \mathbb{C}^N$ of size $|\mathcal{X}| = M = 2^m$.

As shown in Fig. 2.1 the information bits are first encoded by a convolutional en-

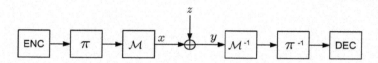

Figure 2.1: Block diagram of BICM transmission: encoder ENC, bit interleaver π, modulator $\mathcal{M}$, metric computing device $\mathcal{M}^{-1}$, deinterleaver π^{-1} and decoder DEC

coder and then a pseudo-random bit interleaver permutes the time index i of the coded bits in order to decorrelate the bits associated with the given symbol and to disperse the burst errors. Every m coded and interleaved bits are then mapped into a M-ary QAM symbol. It is possible to employ any two-dimensional constellation but in [2] was shown that for Gray encoded BICM systems, the calculation complexity for each bit Log Likelihood Ratios (LLR) can be drastically reduced without compromising the system performance [12].

Gray mapping was first proposed by Frank Gray and is defined in [4] as follows:

Let $\mathcal{X}$ denote a signal set of size $M = 2^m$, with minimum Euclidean distance d_{min}. A binary map $\mu : \{0,1\}^m \longrightarrow \mathcal{X}$ is a Gray labelling for $\mathcal{X}$ if, for all

14

$i = 1, \ldots, m$ and $b \in \{0, 1\}$, each $\boldsymbol{x} \in \mathcal{X}_b^i$ has at most one $\boldsymbol{z} \in \mathcal{X}_{\bar{b}}^i$ at distance d_{min}.

In other words, it means that in Gray mapping, adjacent constellation points differ only in a single bit (Fig. 2.2), which reduces the mean square error by one bit error.

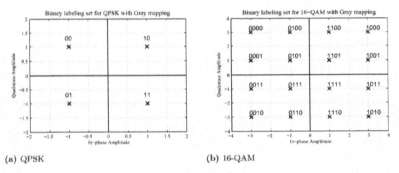

(a) QPSK

(b) 16-QAM

Figure 2.2: Binary labelling sets for QPSK and 16-QAM

On the receiver side, there are a metric computing device $\mathcal{M}^{-1}$, de-interleaver π^{-1} and a soft decision Viterbi convolutional decoder, as suggested by Zehavi in [23].
The main task of the outer[1] receiver is to find the most probable code bit $\hat{u}_i$ for each bit position i, $i = 1, \ldots, m$, within the k-th frequency tone, given a received signal vector $\boldsymbol{y}$, which can be obtained by

$$\hat{u}_i = \arg \left(\max_{b \in \{0,1\}} P(u_i = b | \boldsymbol{y}) \right) \qquad (2.1)$$

This detection rule is referred to as maximum a posteriori probability (MAP) criterion and it gives the optimal reliability information for a certain bit position i of the received vector $\boldsymbol{y}$.

---

[1]A receiver can be divided in an inner and outer part. The transmission parameters such as carrier frequency offset, sampling clock offset, timing synchronization, I/Q-imbalance and multipath channel are estimated and compensated in the inner receiver. The compensated signal is then used to decode the transmitted data in the outer receiver, which mainly consists, in the case of a BICM based OFDM system, of a soft-bit computing device and a Viterbi decoder

Since the outer receiver has to take into account the memoryless binary propagation channel H, which is characterized by an output alphabet $\mathcal{Y}$ and the conditional probability[2] $p(y|u_i = b)$, $y \in \mathcal{Y}$, a maximum likelihood (ML[3]) decoder would be more appropriate in this case, which can be straightforward obtained from (2.1) by using Bayes' theorem.

Now, assuming $P(b)$ to be equal for all possible values of b, which is a very fair assumption, since the transmitted bits are randomized by means of a a pseudo-random scrambler and interleaver (see [21] and [15]), MAP and ML criterion become equivalent:

$$\arg\left(\max_{b\in\{0,1\}} P(u_i = b|\boldsymbol{y})\right) \sim \arg\left(\max_{b\in\{0,1\}} p(\boldsymbol{y}|u_i = b)\right) \tag{2.2}$$

2.1.2 Low Complexity LLR Metrics for BICM Receivers

In this section, a low complexity LLR metric calculation is derived when optimum (unquantized) soft decoding at the receiver is employed. This type of receiver is also referred as interference ignorant receiver.

Let us consider the simplest possible scenario, which is a random BPSK symbol transmission over an AWGN channel. As we know, BPSK modulation can be defined as $x = (-1)^u$, where $u \in \{0, 1\}$ and $x \in \{\pm 1\}$.

Thus, the probability distribution function for the received symbol y is given by

$$p(y|x = \pm 1) = \frac{1}{\sqrt{2\pi\sigma^2}} e^{-\frac{1}{2\sigma^2}(y \mp 1)^2} \tag{2.3}$$

Instead of probability and for simplicity's sake we can use the ML metric

$$\begin{aligned}
\lambda(y, x) &= \max_{x\in\{\pm 1\}} \ln p(y|x) \\
&= \max\left\{-\frac{1}{2}\ln(2\pi\sigma^2) - \frac{1}{2\sigma^2}|y - x|^2\right\} \\
&= \min\left\{\frac{1}{N_0}|y - x|^2\right\}
\end{aligned} \tag{2.4}$$

Actually, ML metric involves summation instead of maximum operation and the above metric is commonly referred as max log MAP metric [12].

---

[2]Capital $P(x)$ and lower-case $p(x)$ are used for discrete probabilities and continuous probabilities, respectively

[3]ML decoder maximizes Prob(received|transmitted).

MAP decoder maximizes Prob(transmitted|received).

So, max log MAP metric is equivalent to pick up the symbol $\hat{x}$ with the least Euclidean distance to the received symbol y, $D(y, x) = ||y - x||^2$:

$$\hat{x} = \arg \left(\min_{x \in \{\pm 1\}} D(y, x) \right) \tag{2.5}$$

Since the decision is a comparison problem between two possible log-likelihoods $\ln p(y|x = +1)$ and $\ln p(y|x = -1)$, a ratio would be more suitable for the solution

$$
\begin{aligned}
\Lambda(y, x) &= \ln \left(\frac{p(y|x = +1)}{p(y|x = -1)} \right) \\
&= \ln(p(y|x = +1)) - \ln(p(y|x = -1))
\end{aligned} \tag{2.6}
$$

where $\Lambda(y, x) \in \mathbb{R}$ is the so-called log-likelihood ratio.

The sign of $\Lambda(y, x)$ corresponds to its hard decided binary value. A positive LLR indicates that the bit is more likely to be 0, a negative LLR indicates that the bit $\hat{u}$ is more likely to be 1 and the magnitude $\Lambda(y, x)$ indicates how sure we are about the hard decision of the decoded bit. But the decision will be first taken in the soft Viterbi decoder. Thus, the data is passed from one component to another in form of LLRs. Clearly, practical digital implementations can only use a fixed point approximations of the real numbers (LLR quantization), which is obviously not a trivial task [10, 9] and it is out of the scope of this thesis.

Obviously, the BPSK example was so simple that is not possible to talk about a simplified LLR. So, let us consider a higher modulation scheme such as a Gray coded QPSK constellation (Fig. 2.2a), where $x[k] = x_I[k] + jx_Q[k]$ denote the transmitted symbol on the k-th subcarrier over a flat-fading real valued channel. The received symbol is then $y_k = h_k x_k + z_k$.

So, formulating the max log MAP metric from (2.4) in terms of real and imaginary parts and taking the channel into consideration, we obtain:

$$
\begin{aligned}
\lambda_b^i(y, h, x) &= \min_{x \in \mathcal{X}_b^i} \left\{ (y_R - hx_R)^2 + (y_I - hx_I)^2 \right\} \\
&= \min_{x \in \mathcal{X}_b^i} \left\{ |y|^2 - 2\bar{y}_R x_R - 2\bar{y}_I x_I + |h|^2 x_R^2 + |h|^2 x_I^2 \right\}
\end{aligned} \tag{2.7}
$$

where $\bar{y}_k = h_k^* y_k$ is the matched filter output, the subscripts $(\cdot)_R$ and $(\cdot)_I$ indicate the real and imaginary part of the complex symbol, respectively and since $|y|^2$ is common for all λ, it can be omitted in the computation.

Using (2.7) in (2.6), we obtain:

$$
\begin{aligned}
\Lambda_i &= \lambda_1^i(y,h,x) - \lambda_0^i(y,h,x) \\
&= \max_{x \in \mathcal{X}_1^i} \left\{ 2\bar{y}_R x_R + 2\bar{y}_I x_I - |h|^2 x_R^2 - |h|^2 x_I^2 \right\} \\
&\quad - \max_{x \in \mathcal{X}_0^i} \left\{ 2\bar{y}_R x_R + 2\bar{y}_I x_I - |h|^2 x_R^2 - |h|^2 x_I^2 \right\}
\end{aligned}
\tag{2.8}
$$

Furthermore, we can decouple the real and imaginary parts of (2.8):

$$
\begin{aligned}
\Lambda_i &= \max_{x_R \in \mathcal{X}_1^i} \left\{ 2\bar{y}_R x_R - |h|^2 x_R^2 \right\} + \max_{x_I \in \mathcal{X}_1^i} \left\{ 2\bar{y}_I x_I - |h|^2 x_I^2 \right\} \\
&\quad - \max_{x_R \in \mathcal{X}_0^i} \left\{ 2\bar{y}_R x_R - |h|^2 x_R^2 \right\} - \max_{x_I \in \mathcal{X}_0^i} \left\{ 2\bar{y}_I x_I - |h|^2 x_I^2 \right\}
\end{aligned}
\tag{2.9}
$$

From Fig. 2.2a we can also see that when the bit x_1 in the QPSK constellation toggles from $1 \rightleftharpoons 0$ only the real part of the constellation is affected, and for the bit x_2 only the imaginary part. And if we look further, in the 16-QAM or even higher Gray mapped modulation schemes we conclude that

$$
\begin{aligned}
\Lambda_{i'} &= \max_{x_R \in \mathcal{X}_1^{i'}} \left\{ 2\bar{y}_R x_R - |h|^2 x_R^2 \right\} - \max_{x_R \in \mathcal{X}_0^{i'}} \left\{ 2\bar{y}_R x_R - |h|^2 x_R^2 \right\} \\
&\hspace{5cm} i' = 1, \ldots, \frac{m}{2} \\
\Lambda_{i''} &= \max_{x_I \in \mathcal{X}_1^{i''}} \left\{ 2\bar{y}_I x_I - |h|^2 x_I^2 \right\} - \max_{x_I \in \mathcal{X}_0^{i''}} \left\{ 2\bar{y}_I x_I - |h|^2 x_I^2 \right\} \\
&\hspace{5cm} i'' = \frac{m}{2} + 1, \ldots, m
\end{aligned}
\tag{2.10}
$$

From (2.10), the LLR of the first bit for QPSK with a normalization factor of $\frac{\sigma}{\sqrt{2}}$ is given as

$$
\begin{aligned}
\Lambda_1^{\mathrm{QPSK}} &= \max_{x_R = \frac{\sigma}{\sqrt{2}}} \left\{ 2\bar{y}_R x_R - |h|^2 x_R^2 \right\} - \max_{x_R = \frac{\sigma}{\sqrt{2}}} \left\{ 2\bar{y}_R x_R - |h|^2 x_R^2 \right\} \\
&= -2\sqrt{2}\sigma \bar{y}_R
\end{aligned}
\tag{2.11}
$$

and the LLR of the second bit is given as

$$
\begin{aligned}
\Lambda_2^{\mathrm{QPSK}} &= \max_{x_I = -\frac{\sigma}{\sqrt{2}}} \left\{ 2\bar{y}_I x_I - |h|^2 x_I^2 \right\} - \max_{x_I = \frac{\sigma}{\sqrt{2}}} \left\{ 2\bar{y}_I x_I - |h|^2 x_I^2 \right\} \\
&= -2\sqrt{2}\sigma \bar{y}_I
\end{aligned}
\tag{2.12}
$$

Most of the simplifications occur just with replacing $\{x_{R,I}\}$ with its actual value $\{\pm \frac{\sigma}{\sqrt{2}}\}$.

2.2 Interference-Aware System

In the last section a simple and general expression of max log MAP metric computation in a flat fading channel, without any co-channel interference was introduced. In this section we will go further and consider a more realistic scenario, which takes into account co-channel interferences between hexagonal cells. Co-channel interference occurs when, a MS simultaneously receives signals from the serving eNodeB , as well as from co-channel neighbour eNodeB. Indeed, the downlink (DL) performance of a cellular system is strongly limited by co-channel interferences, especially in the cellular systems with a frequency reuse of factor one or fractional frequency reuse systems [11].

In most cases, there are a maximum of two dominant interferers, one if MS is close to the

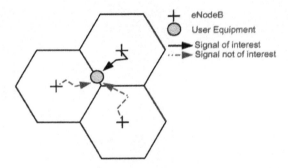

Figure 2.3: Co-channel interference between hexagonal cells with a frequency reuse of factor one

cell boundaries, two near cell edges. For simplicity sake and without lose of generality, only one interferer in this work is taken in account.

2.2.1 IA Receiver: System & Signal Model

We consider a single frequency reuse cellular network, with two neighbour cells, which use a BICM based OFDM system for DL transmission and one IA-receiver, as shown in Fig. 2.3 One signal of interest and one interferer. We also assume that the CP is of appropriate length, the BS's are synchronized for transmissions, the IA-R knows about the modulation scheme of the interferer and can also estimate its propagation channel.

Cascading the IFFT at the BS and the FFT at the MS with CP extension, transmission at the $k - th$ subcarrier can be then expressed as follow:

$$\begin{aligned} \mathbf{y}_k &= \mathbf{h}_{1,k} x_{1,k} + \mathbf{h}_{2,k} x_{2,k} + \mathbf{z}_k \\ &= \mathbf{H}_k \mathbf{x}_k + \mathbf{z}_k \ , \quad k = 1, 2, \ldots, K \end{aligned} \tag{2.13}$$

where K is the total number of subcarriers, $\mathbf{H}_k = [\mathbf{h}_{1,k} \ \mathbf{h}_{2,k}]$ is the virtual channel from two BSs to the user at the $k - th$ frequency tone, $\mathbf{x}_k = [x_{1,k} \ x_{2,k}]^T$ and $\mathbf{z}_k$ is the noise vector with $z_k \in \mathcal{CN}(0, \sigma^2)$. Each subcarrier corresponds to a symbol from a constellation map $x_1 \in \mathcal{X}_1$ and $x_2 \in \mathcal{X}_2$.

The max-log MAP bit metric for bit b of the signal of interest x_1 can be then obtained by inserting the new received signal model $\mathbf{y}_k$ from (2.13) in (2.6)

$$\begin{aligned} \lambda_{1,b}^i(\mathbf{y}, \mathbf{H}, \mathbf{x}) &= \min_{\substack{x_1 \in \mathcal{X}_{1,b}^i \\ x_2 \in \mathcal{X}_2}} \left\{ ||\mathbf{y} - \mathbf{h}_1 x_1 - \mathbf{h}_2 x_2||^2 \right\} \\ &= \min_{\substack{x_1 \in \mathcal{X}_{1,b}^i \\ x_2 \in \mathcal{X}_2}} \left\{ ||\mathbf{y}||^2 + ||\mathbf{h}_1 x_1||^2 - 2\left(\bar{y}_1 x_1^*\right)_R + 2(\rho_{12} x_1^* x_2)_R \right. \\ &\qquad\qquad \left. -2(\bar{y}_2 x_2^*)_R + ||\mathbf{h}_2 x_2||^2 \right\} \end{aligned} \tag{2.14}$$

where $\bar{y}_1 = \mathbf{h}_1^H \mathbf{y}$, and $\bar{y}_2 = \mathbf{h}_2^H \mathbf{y}$ are the output of the MF and $\rho_{12} = \mathbf{h}_1^H \mathbf{h}_2$ is the correlation coefficient between the two channels. Splitting (2.14) into real and imaginary parts and omitting the common term $||\mathbf{y}||^2$ we have

$$\begin{aligned} \lambda_{1,b}^i(\mathbf{y}, \mathbf{H}, \mathbf{x}) &= \min_{x_1 \in \mathcal{X}_{1,b}^i \ x_2 \in \mathcal{X}_2} \left\{ ||\mathbf{h}_1||^2 |x_1|^2 + ||\mathbf{h}_2||^2 |x_2|^2 - 2\left(\bar{y}_1 x_1^*\right)_R \right. \\ &\qquad + \left(2\left(\rho_{12,R} x_{1,R} + \rho_{12,I} x_{1,I}\right) - 2 y_{2,R} \right) x_{2,R} \\ &\qquad \left. + \left(2\left(\rho_{12,R} x_{1,I} - \rho_{12,I} x_{1,R}\right) - 2 y_{2,I} \right) x_{2,I} \right\} \end{aligned} \tag{2.15}$$

Thus, if $|\mathcal{X}_1| = M$ and $|\mathcal{X}_2| = M'$, we need to compare MM' possible values and take the minimum between them, which can be by high modulation schemes very inefficient in terms of performance and computational complexity.

Lets define

$$\begin{aligned} \eta_1 &= \rho_{12,R} x_{1,R} + \rho_{12,I} x_{1,I} - y_{2,R} \\ \eta_2 &= \rho_{12,R} x_{1,I} - \rho_{12,I} x_{1,R} - y_{2,I} \end{aligned} \tag{2.16}$$

(2.15) can be minimized if $x_{2,R}$ and $x_{2,I}$ are in the opposite directions (have the opposite signs) of η_1 and η_2, respectively.

Let $\mathcal{X}_1$ and $\mathcal{X}_2$ be each a normalized symbol set with a Gray mapped QPSK constellation. So, $|x_1|^2 = |x_2|^2 = 1$ and since $(||\mathbf{h}_1||^2 + ||\mathbf{h}_2||^2)$ and the factor 2 are common for both max log MAP metrics λ^i with $i = 1, 2$, they can be neglected.
Hence, (2.15) can be written as

$$\lambda_{1,b}^i = \max_{x_1 \in \mathcal{X}_{1,b}^i \ x_2 \in \mathcal{X}_2} \left\{ \bar{y}_{1,R} x_{1,R} + \bar{y}_{1,I} x_{1,I} + |\eta_1||x_{2,R}| + |\eta_2||x_{2,I}| \right\} \qquad (2.17)$$

From (2.17), the LLR for the first bit of the QPSK symbol of interest x_1 is given by

$$\begin{aligned}
\Lambda_1^{\text{QPSK-QPSK}} &= \max_{\substack{x_{1,R}=-\frac{\sigma}{\sqrt{2}} \\ x_{1,I}=\pm\frac{\sigma}{\sqrt{2}}}} \left\{ \bar{y}_{1,I} x_{1,I} + |\eta_1||x_{2,R}| + |\eta_2||x_{2,I}| \right\} \\
&\quad - \max_{\substack{x_{1,R}=\frac{\sigma}{\sqrt{2}} \\ x_{1,I}=\pm\frac{\sigma}{\sqrt{2}}}} \left\{ \bar{y}_{1,I} x_{1,I} + |\eta_1||x_{2,R}| + |\eta_2||x_{2,I}| \right\} - \sqrt{2}\sigma \bar{y}_{1,R}
\end{aligned} \qquad (2.18)$$

and the second bit of the same QPSK symbol is

$$\begin{aligned}
\Lambda_2^{\text{QPSK-QPSK}} &= \max_{\substack{x_{1,R}=-\frac{\sigma}{\sqrt{2}} \\ x_{1,I}=\pm\frac{\sigma}{\sqrt{2}}}} \left\{ \bar{y}_{1,R} x_{1,R} + |\eta_1||x_{2,R}| + |\eta_2||x_{2,I}| \right\} \\
&\quad - \max_{\substack{x_{1,R}=\frac{\sigma}{\sqrt{2}} \\ x_{1,I}=\pm\frac{\sigma}{\sqrt{2}}}} \left\{ \bar{y}_{1,R} x_{1,R} + |\eta_1||x_{2,R}| + |\eta_2||x_{2,I}| \right\} - \sqrt{2}\sigma \bar{y}_{1,I}
\end{aligned} \qquad (2.19)$$

3 Structure of the Simulator

This chapter is dedicated to the Matlab implementation of a BICM based OFDMA transceiver system in the equivalent baseband. We assume that the time and frequency resources are organized as in the LTE standard, but without considering all the details and only the downlink transmission scheme is considered.

Each implemented block will be explained and discussed. A particular focus here is the performance of the interference aware receiver in an inter-cell interference environment with one dominant interferer. Additional optimization techniques are also proposed and discussed in this chapter such as boosted pilot symbols in the interferer and REs puncturing in the signal of interest.

In fact, LTE standard allows a certain degree of freedom in the signal generation, which makes these two optimization techniques easy to incorporate in real systems.

3.1 System Parameters & General Code Structure

Fig. 3.1 shows the general code structure of the simulation testbed, which is composed of five main building blocks:

- Control unit, which is used to control the behaviour and priorities of the different components of the implementation, some of the parameters are set as default and others are adjustable.

- Serving eNodeB,

- Interferer eNodeB,

- Virtual channel,

- and the UE

Each of these block will be discussed in the following sections, where the interferer eNodeB and the serving eNodeB are both summarized in the transmitter section. The major difference between these two kind of transmitters is notable only if one of the optimization techniques introduced above is considered.

In (3.1), the essential system parameters of a 5MHz DL PHY are illustrated.

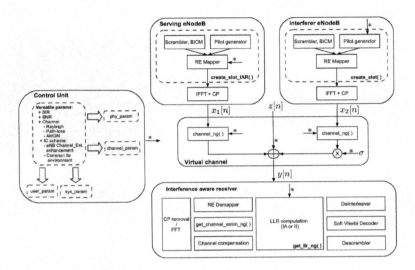

Figure 3.1: Structure of the Simulation Testbed

LTE transmissions are organized into $10ms$ radio frames, which corresponds to 76800 samples by $N_{fft} = 512$. Each of these radio frames is divided into 10 equally sized subframes. Each subframe consists of 2 equally sized slots with $T_{slot} = 0.5ms$. In the frequency domain, one slot consists of 7 OFDM symbols, where each OFDM symbol consists of a certain number of active subcarriers N_c. The REs of the first symbol of the implementation are filled only with pilot symbols, while the REs of the next two symbols are left empty, since the control channels in the implementation are not considered.

Fig. 3.2 illustrates one OFDM slot. In the time domain it consists of 3840 samples with $T_s = 130.21ns$ and in the resource grid illustration it consists of 25 RBs × 7 Symbols. From the illustration of the power spectrum of the generated LTE signal, we can also see the 5MHz bandwidth as well as the sampling rate.

In the simulation testbed only one slot will be generated, sent and processed in the context of a Monte Carlo simulation.

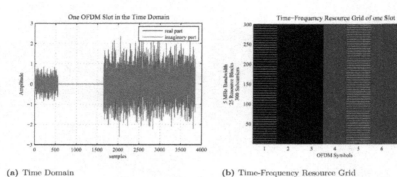

(a) Time Domain

(b) Time-Frequency Resource Grid
(Black: empty REs. White: Pilots. Gray: Data)

Figure 3.2: Illustration of one LTE Slot

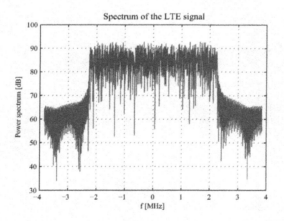

Figure 3.3: Power Spectrum of LTE DL Signal

Parameter	Value
Bandwidth B	5 MHz
FFT size N_{FFT}	512
Active subcarriers N_c	300
Subcarrier spacing Δf_c	15 kHz
Sampling rate f_s	$\Delta f_c \times N_{FFT} = 7.68$ MHz
Frame duration	10 ms
Subframe/Slot length	14/7 OFDM symbols, 1/0.5 ms, 7680/3840 samples
REs per RB	12
Coding type	convolutional code R = 1/2, 3/4 punctured
Generator polynomial	$(7,5)_8$, $(133,171)_8$
Data modulation	QPSK,Gray Mapping
Transmission mode	SISO (TM1)

Table 3.1: System Parameters

3.2 Radio Propagation Channel

Two functions have been implemented to generate the radio propagation channels: *channel_ng( )* and *channel( )*. Both functions generate an AWGN channel and Rayleigh fading but the first one takes the interference into account, and is therefore more suitable in generating the virtual channel, while the *channel( )* function is more appropriate for simulating AWGN, path-loss or Rayleigh without considering the interference. They both take the time domain baseband LTE signal and the transport block size as input. The convolution between the transmitted signal and the channel happens inside these function, so as output we get the signal after passing through the channel and also the channel itself in the time and frequency domain.

3.2.1 Simulation of AWGN-Channel model

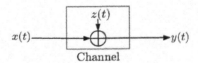

Figure 3.4: AWGN Channel model

All transmission channels have the superposition of the so-called additive white Gaussian noise (AWGN) at the input of the receiver in common. It is an idealized conceptual model for thermal noise, with a constant spectral density and a normal distribution of amplitude. It is characterized by two parameters, μ (mean) and σ^2 (variance) : $\mathcal{N}(\mu = 0, \sigma^2)$ and has the following density function:

$$p(z) = \frac{1}{\sqrt{2\pi}\sigma} e^{-\frac{z^2}{2\sigma^2}} \tag{3.1}$$

In Fig. 3.5a we can see that the generated AWGN fits the Gaussian distribution almost

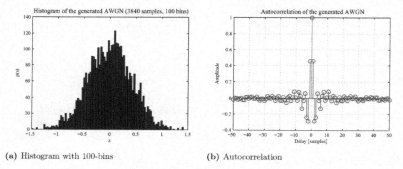

(a) Histogram with 100-bins   (b) Autocorrelation

Figure 3.5: generated AWGN, where $z \in \mathcal{CN}(0, 0.35)$

perfectly and Fig. 3.5b shows that the samples of the generated Gaussian white noise are almost completely uncorrelated after $0.5\mu s$. Thus, the received signal $y(t)$ passed through the simulated AWGN channel model (Fig. 3.4) can be represented as

$$y(t) = x(t) + z(t) ; \tag{3.2}$$

where $z(t)$ is the zero mean circularly symmetric complex white Gaussian noise of variance N_0. The SNR is then defined as:

$$SNR = \frac{E\left\{|x(t)|^2\right\}}{E\left\{|z(t)|^2\right\}} = \frac{P_s}{\sigma_z^2} \tag{3.3}$$

where P_s denotes the average power per symbol. Fig. 3.6 gives a qualitative impression of the impact of SNR on one received OFDM symbol just after the CP removal and the FFT.

The simulation of the AWGN channel plays an important role in the validation of the

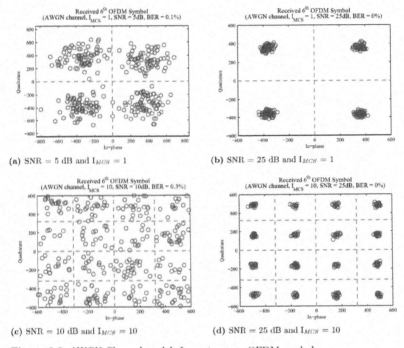

(a) SNR = 5 dB and I_{MCS} = 1

(b) SNR = 25 dB and I_{MCS} = 1

(c) SNR = 10 dB and I_{MCS} = 10

(d) SNR = 25 dB and I_{MCS} = 10

Figure 3.6: AWGN-Channel model: Impact on one OFDM symbol

implemented transceiver, specifically in the validation of all the sender-receiver signal processing blocks apart from the channel estimator and equalizer blocks, these can be validated with other channel models such as the Rayleigh channel.

The theoretical bit error probability in the case of QPSK is given by:

$$P_b = \frac{1}{2} erfc(\sqrt{2\gamma_b}) : \tag{3.4}$$

where $\gamma_b := E_b/N_0$ is the *SNR* per bit and $erfc(\alpha) := \frac{2}{\sqrt{\pi}} \int_0^{\alpha} e^{-x^2} dx$ is the error function. Now, any difference between the simulated bit error rate and the theoretical BER means an error somewhere in the code. When adding the required noise, it is important to consider the additional overheads in an OFDM symbol in both time and frequency domain such as CP, zero-padded frequencies and control channels. For instance, if we consider an LTE system and if the *SNR* per symbol is already given we can determine the γ_b as given below:

$$\gamma_b = \left(\frac{N_{fft} + N_{cp} + \left(\frac{N_{cp_0} - N_{cp}}{N_{symb}} \right)}{N_{fft}} \right) \left(\frac{N_{sc} N_{symb}}{TBS} \right) \frac{E_s}{N_0} \tag{3.5}$$

From Fig. 3.7 we can conclude that a big part of the implemented transceiver is validated.

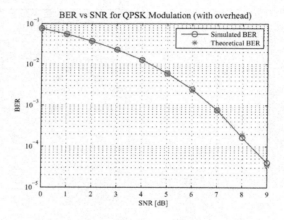

Figure 3.7: BER theoretical vs simulated

3.2.2 Path Loss Channel Model

The initial planning of any Radio Access Network begins with a Radio Link Budget. It is used to estimate the maximum allowed path loss and the corresponding cell range and it takes into account the gains and losses from the transmitter, through the medium to the receiver.

The path loss of a channel *PL* describes an attenuation of the signal power along the distance between the serving BS and the UE and is defined as the ratio of the sender signal power P_{TX} to the received signal power P_{RX}

$$PL\,|_{dB} = 10\log_{10}\left(\frac{P_{TX}}{P_{RX}}\right) \tag{3.6}$$

The relation between P_{RX} and P_{TX} can be expressed with the Friis transmission equation:

$$P_{RX}(d) = \left(\frac{\lambda}{4\pi d}\right)^2 G_{TX}\,G_{RX}\,P_{TX} \tag{3.7}$$

where d is the radio path length. G_{TX} and G_{RX} are the antenna gains at the transmitter and receiver. f_c is the carrier frequency of the signal and λ is the wavelength. The inverse of the factor in parentheses is the so-called free-space path loss. From the above formula we can see that the received power decreases proportionally to the square of the separation distance d and the carrier frequency $f_c = \frac{c}{\lambda}$, where c is the light propagation velocity.

Thus, from Eq.(3.6) and (3.7) we obtain:

$$PL\,|_{dB} = 20\log_{10}(d) + 20\log_{10}(f_c) - 20\log_{10}\left(\frac{c}{4\pi}\right) - G_{TX}|_{dBi} - G_{RX}|_{dBi} \tag{3.8}$$

There are other empirically-based path loss models such as IMT-Advanced, which was approved by the ITU-R and the geometry based SCM, developed jointly by 3GPP and 3GPP2. Since these models cover numerous special cases such as outdoor-to-indoor , indoor-to-outdoor or even outdoor-to-vehicle situations (chapter 20 in [16]), they are used for the evaluation of the performance of new LTE devices.

3.2.3 Simulation of Rayleigh Fading Channel

In this section we will introduce the Rayleigh fading channel. It is also an idealized channel model, which is very important in the validation of the implemented channel estimator, interpolation and the equalizer.

The Rayleigh channel model is a multipath channel with non line of sight (NLOS) propagation properties 3.9.

It is characterised by its CIR length (number of taps), a Rayleigh distributed complex gain (Fig. 3.10) and the delays associated with the taps.

Hence, the distribution of the magnitude of its time-varying channel impulse $|h(t, \tau)|$,

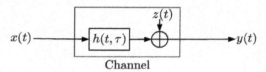

Channel

Figure 3.8: Rayleigh Channel Model

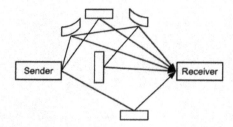

Figure 3.9: Non Line of Sight (NLOS)- Multipath

where the variable t represents the time-variation and τ the channel multipath delay for a fixed t, can be defined as:

$$p(|h|) = \frac{|h|}{\sigma^2} e^{-\frac{|h|^2}{2\sigma^2}}$$

$$p(\theta) = \frac{1}{2\pi} , \quad -\pi \leqslant \theta \leqslant \pi$$

(3.9)

(3.9) means that the phases of the multipath received signal are uniformly distributed and its in-phases and quadrature components are two independent Gaussian random variables having equal variance σ^2.

From Fig. 3.11 we can see the importance of the channel estimator and correction in a Rayleigh channel. A receiver without channel equalization in a multipath channel is incapable to work correctly even in the best-case input SNR.

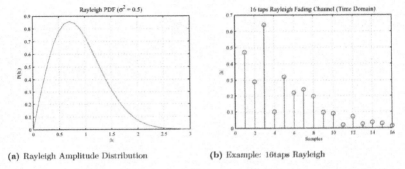

(a) Rayleigh Amplitude Distribution   (b) Example: 16taps Rayleigh

Figure 3.10: Rayleigh Distribution and Example

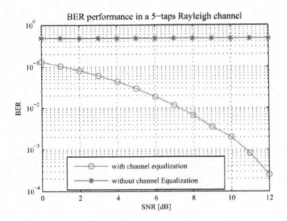

Figure 3.11: BER performance in a Rayleigh Propagation Channel

3.3 The Baseband Part of the Transmitter

Fig. 3.12 depicts the block diagram of the simulated transmitter. First of all a stream of random bits is generated, then the transmitter starts encoding the information bit stream with a convolutional code of rate R for forward error correction (FEC). After interleaving, the permuted coded bits are scrambled and Gray-mapped. We distinguish between three main types of data: pilots, control data and information data. Once each of these data type is mapped to its predefined position within the time-frequency resource grid, each resource element is modulated into an OFDM subcarrier (Fig. 3.13).

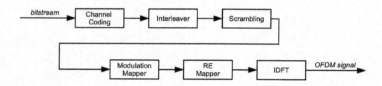

Figure 3.12: Block Diagram of the Baseband Transmitter

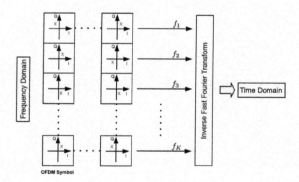

Figure 3.13: OFDM Modulation, OFDM Symbol

3.3.1 Convolutional Coding and Puncturing

The convolutional coding is a forward error correction technique, used in order to improve the capacity of a noisy channel by adding some redundant information to the input data. The encoder can be realized as a sequential circuit with k inputs, n outputs and a number of memory elements m. So, the convolutional codes can be specified by $(n, k, [m])$ [3]

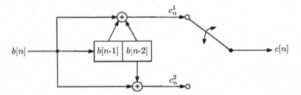

Figure 3.14: Encoder with generator polynomials $g_1 = [111]$ and $g_2 = [101]$

or also by its code rate $R = k/n$ and the so-called constraint length $K = k \times (m-1)$ or even just by its generator polynomials. Generator polynomials define the connection structure between the memory elements, the number of generator sequences, defines the number of XOR elements, which also means the number of outputs n and the number of memory elements m is:

$$m = \max_{1 \le \nu \le n} \deg g_\nu(D) \tag{3.10}$$

Table 3.2 shows some optimum convolutional codes with their generator polynomials for code rate $R = 1/2$ and number of memory elements $m = 2 \ldots 6$.

Each coded bit is a function of the input bit and the state of the memory elements.

$g_1(D)$	$g_2(D)$
$1 + D + D^2$	$1 + D^2$
$1 + D + D^3$	$1 + D + D^2 + D^3$
$1 + D^2 + D^3 + D^3 + D^5 + D^6$	$1 + D + D^2 + D^3 + D^6$

Table 3.2: Generator polynomials for $R = 1/2$ with $m = 2 \ldots 6$

With every new bit the state of the shift register changes resulting in n new encoded output bits. A very convenient way for describing the behaviour of the encoder is the trellis diagram.

Trellis diagram consists of nodes and branches. Each node represents one of the states, that can be taken by the memory elements of the encoder. Branches notify a state transition, depending on the input data bit.

The corresponding trellis diagram of the encoder in Fig. 3.14 is shown in Fig. 3.15, where the dashed line represents a 1 and the solid line a 0.

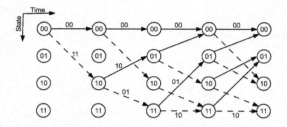

Figure 3.15: Trellis diagram of $(7,5)_8$ encoder

Now to get higher rates from a rate-1/2 mother code, some encoded bits are just deleted and not transmitted. This simple but very efficient technique is called puncturing. In fact, puncturing allows us to produce codes of many different rates dynamically without increasing the complexity of the decoding algorithm, which only depends on the mother code. For instance if we want to get a 3/4-code rate we need to puncture 2 bits every 6 encoded bits, which means a total code rate of $\frac{1/2}{4/6} = 3/4$.

3.3.2 Bit-Interleaver

Errors due to the multipath fading of radio propagation channels tend to occur burst-wise. Thus, the ability of FEC decoders to correct errors without a retransmission request is lost.

A good way to deal with these kind of errors is to disperse a burst error into different FEC blocks, which can be realized by mean of interleaver.

Bit-interleaver is a device, that permutes the ordering of the input bits. Proportionally to its depth, bit errors per FEC block are reduced but on the other hand, buffer requirement increases. Hence, an ideal interleaver would be a random permutation of the bit order over all input bits. But considering the additional introduced processing delay in both the transmitter and the receiver, a practical application of bit-interleaver requires a short interleaving depth.

In the function *do_interlv()* are two types of interleaver implemented, an ideal interleaver and the LTE quadratic permutation polynomials (QPP) interleaver [TS 136.212v10] with an interleaving relationship of the form:

$$\pi(i) = (f_1 i + f_2 i^2) \mod K, \tag{3.11}$$

where K is the input block size, i the output index and f_1, f_2 are some permutation coefficients.

QPP interleaver are contention-free interleaver, which means they are very flexible in supported parallelism. This feature make them very attractive.

3.3.3 Bit-Level Scrambling

Scrambling is a modulo 2 multiplication (XOR) between the input coded bits and a 32-bits gold sequence. The scrambling sequence depends on the PHY cell identity and therefore differs between neighbouring cells. When the receiver descrambles a received bitstream with a known cell specific scrambling sequence, interference from other cells will be descrambled incorrectly and therefore only appear as uncorrelated noise.

The concatenation of scrambler and interleaver aims to randomize the interfering signals at the receiver side and thus transforms intercell interference ignoring to a kind of interference suppression process, since the interferer can be considered as a part of the AWGN.

3.4 Interference Model

The capacity of a cellular network is limited by two main factors: noise and interference, which can be characterized by the signal to noise and interference ratio ($SNIR$):

$$SNIR = \frac{S}{N + I} \; , \tag{3.12}$$

with S the signal of interest power, I the sum of all interfering signals and N the AWGN power. By densely populated areas is interference the most limiting factor. Thus, the noise can be ignored and instead of $SNIR$, we talk about signal-to-interference ratio (SIR):

$$SIR = \frac{S}{I} \; , \tag{3.13}$$

We can divide interference into two major classes, intra-cell and inter-cell interference. The first refers to all interference sources from the same cell. These are easier to avoid or mitigate. The latter refers to interference between sources from different cells.
In this work we consider an inter-cell interference in an environment without any frequency reuse restrictions.

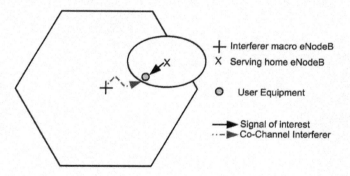

Figure 3.16: Interference Model

Fig. 3.16 shows the simulated interference model. The difference of the geometric form of these two cells emphasizes the significant power imbalance between the home eNodeB and the macro eNodeB.
For simplicity sake we consider just one interfering symbol x_2 and one symbol of interest

x_1, sent through the propagation channel h_2 and h_1, respectively. Both are QPSK modulated and we also suppose that the BSs are perfectly synchronized.

Cascading the IFFT at the BS and the FFT at the MS received signal model at the k-th subcarrier can be then expressed as follows:

$$y = H_1 x_1 + H_2 x_2 + z \tag{3.14}$$

with $z \in \mathcal{CN}(0, \sigma^2)$.

Depending on which type of receiver is used: interference aware or interference ignorant receiver, y can be interpreted differently. Interference aware receiver makes use of the structure of the interfering symbol. It supposes, it has a specific modulation with a specific mapping and also needs knowledge about the position of the pilots of the interfering signal, while the II-R makes use of the randomization of the interference and just considers it as part of the AWGN.

This difference in the outer receiver input signal interpretation has its first noticeable impact on the metric computing device (Fig. 2.1).

3.5 The Baseband Part of the Receiver

Basically, the receiver performs the reverse operations of the transmitter with some supplementary operations. In this simulation, it is assumed that the transmitter and the receiver are perfectly synchronized.

The overall receiver structure is shown in Fig. 3.17. Firstly, the CP is removed and an OFDM demodulation using the FFT is performed. After the channel transfer function estimation, the complex modulated signal is equalized.

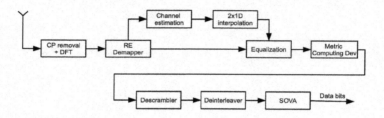

Figure 3.17: Baseband Part of the Receiver

3.5.1 Pilot-based Channel Estimation

In order to work correctly, a receiver needs to equalize the effect of multipath channel on the received signal 3.2.3. Equalization supposes accurate knowledge of the propagation channel, otherwise it could have the opposite desired effect.

Now to get the accurate channel estimation, the transmitter will add some reference symbols in certain positions well known by the receiver. These reference symbols are named pilots and are generated from a pseudo-random sequence, with different initializations for each cell-ID.

These pilots are inserted within the first and the third last OFDM symbol of each slot with a frequency spacing of six subcarriers (Fig. 3.18). The pilot based channel estimation allows, in comparison to the preamble based channel estimation, the assumption that the channel might be a linear time-variant. However, it is assumed that the channel during the transmission of one OFDM symbol is invariant. This actually depends on the speed of the UE and the speed of the surrounding objects.

The channel estimation happens in the frequency domain and it is generally a division of

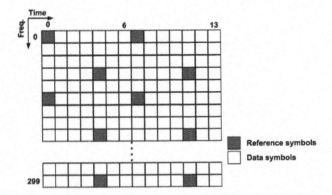

Figure 3.18: Pilots within a Subframe

the received REs containing the pilots with the used pilots on the transmitter side. The UE needs only to know the position of the pilots within the signal, which can be directly determined from the Cell-ID, and their modulation scheme. But since the reference signals in the LTE specification are normalized QPSK symbols, we use only a multiplication of the complex conjugate version of the used pilots with the REs containing the CS-RS.

$$
\begin{aligned}
\hat{H}_{n_p,k_p} &= Y_{n_p,k_p} X^*_{n_p,k_p} \\
&= (H_{n_p,k_p} X_{n_p,k_p} + Z_{n_p,k_p}) X^*_{n_p,k_p} \\
&= H_{n_p,k_p} \underbrace{|X_{n_p,k_p}|^2}_{1} + \tilde{Z}_{n_p,k_p} \ ;
\end{aligned}
\tag{3.15}
$$

where n_p and k_p denote the symbol and subcarrier index of the p-th pilot, respectively, $\tilde{Z}_{n_p,k_p}$ is a complex scaled frequency-domain noise and since the pilots are normalized QPSK symbols, $|X_{n_p,k_p}|^2 = 1$.

Now, to determine the quality of channel estimations and especially the quality of the used interpolation technique and also to compare different estimation and interpolation methods with each other, we calculate the so-called mean square error between the ideal and the estimated channel transfer function as complement for the BER, which can be used for every quality and performance investigation.

Let $\varepsilon_{n,k}$ be the error between the actual CTF $H_{n,k}$ and the estimated CTF $\hat{H}_{n,k}$:

$$
\varepsilon_{n,k} = H_{n,k} - \hat{H}_{n,k}
\tag{3.16}
$$

Consequently, the mean square error can be obtained by:

$$MSE|_{dB} = E\left\{|\varepsilon_{n,k}|^2\right\}|_{dB} \tag{3.17}$$

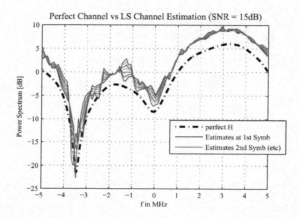

Figure 3.19: Channel Estimate of a 16-taps Rayleigh Channel by SNR $= 15$ dB

The simplest method to obtain all channel estimates is the $2 \times$ 1-D linear interpolation. Let N_f be the distance between two pilots in the frequency domain and N_p the distance between two OFDM symbols containing the pilots. first of all we interpolate the frequency-domain

$$\hat{H}_{n,k} = \frac{q}{N_f}\left(\hat{H}_{n,(k_p+N_f)} - \hat{H}_{n,k_p}\right) + \hat{H}_{n,k_p} \; ; \text{with } q = 0,\ldots,N_f - 1 \tag{3.18}$$

and then in the time-domain:

$$\hat{H}_{n,k} = \frac{q}{N_t}\left(\hat{H}_{(n_p+N_t),k} - \hat{H}_{n_p,k}\right) + \hat{H}_{n_p,k} \; ; \text{with } q = 0,\ldots,N_t - 1 \tag{3.19}$$

It is also important to notice that in the case of recovering the missing channel information that are not between two pilots we are not talking about interpolation any more but extrapolation.

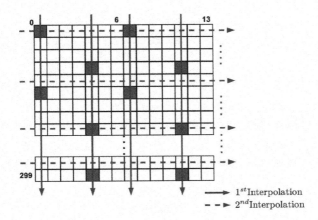

Figure 3.20: 2 × 1-D Interpolation/Extrapolation

3.5.2 De-Puncturing and Soft Output Viterbi Decoding

De-puncturing is the reverse of the puncturing process at the sender side and happens just before the Viterbi decoding. It adds some dummy bits, generally 0 at the punctured bits. Therefor, the used puncturing pattern at the receiver is indispensable. This can be derived from the code rate. The implemented function for this purpose is the *do_depuncturing( )*, which takes as input the interleaved coded metrics and the code rate.

Although convolutional encoding is a quite simple technique, decoding of convolutional codes is much more complicated and especially soft decoding. In fact, soft Viterbi decoding is not just a decoding technique like hard bit decoding. Soft Viterbi algorithm (SOVA) provides also an estimation about the reliability of the decoded sequence [20]. This algorithm has two modification over the classical Viterbi algorithm:

> First, the path metrics used to select the maximum likelihood path through the trellis are modified to take account of a-priori information. Second, the algorithm is modified to provide a soft output for each decoded bit [5].

As SOVA we make use of the existing Matlab function *vitdec( )*.

41

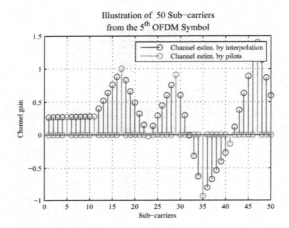

Figure 3.21: Linear Interpolation in the Frequency-Domain

3.5.3 Metric Computing Device

As already mentioned in section 2.1.2, the metric $\Lambda(\bar{y})$ is the LLR of a bit and it can be interpreted as follows:

1. $\Lambda(\bar{y}) > 0 \longrightarrow$ Bit is more likely to be a 0.

2. $\Lambda(\bar{y}) < 0 \longrightarrow$ Bit is more likely to be a 1.

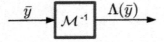

Figure 3.22: Metric Computing Device: Metric as Log Likelihood Ratio

Fig. 3.23 shows an example of $\Lambda(\bar{y})$ in comparison to the actual value of the current bit. The amplitude of the metric can be interpreted as a certainty factor. We can also see that for the bits between position 960 and 965, the absolute value of the amplitude is very low, which is an indication of incertitude or doubt regarding these bits. The decoder is the "decider" of the bit value, between the decoder and the metric computing device data is passed from a component to another in form of LLR. Obviously, the sole noticeable difference between II-R and IA-R is the LLR metric.

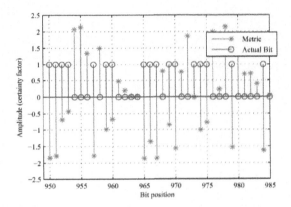

Figure 3.23: LLR vs Bit

The interference ignorant LLR for the first bit $\Lambda_1^{\text{II-R}}$ and the second bit $\Lambda_2^{\text{II-R}}$ defined in (2.11) and (2.12) are implemented as follow:

$$
\begin{aligned}
\Lambda_1^{\text{II-R}} &= \Re\{\bar{y}\} \\
\Lambda_2^{\text{II-R}} &= \Im\{\bar{y}\}
\end{aligned}
\tag{3.20}
$$

The choice of the modulator affects the complexity of the metric computing algorithm. With a Gray mapped constellation, the computing complexity is drastically reduced without compromising the system performance.

It is important to notice that on the DL we do not equalize the amplitude, so the noise variance is constant. Only the phase is corrected by mean of MF, $\bar{y} = \hat{H}^* y$ is the received signal after MF. Actually, an LLR concentrated in ± 1 is as good as hard bit decoding. In Fig. 3.24, 3.25 and 3.26 we can see the effect of a Rayleigh channel on an OFDM symbol, the output of an usual equalizer and the output of an MF, respectively. While the amplitude of an OFDM symbol after MF doesn't change in comparison to the OFDM symbol after the channel, the amplitude of the same OFDM symbol is concentrated on ± 1. The interference aware LLR for the first bit $\Lambda_1^{\text{IA-R}}$ and the second bit $\Lambda_2^{\text{IA-R}}$, assuming

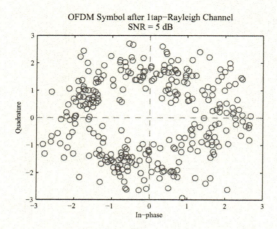

Figure 3.24: 1tap Rayleigh Channel effects on an OFDM Symbol by $SNR = 5$ dB

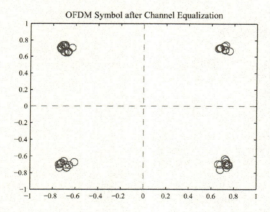

Figure 3.25: OFDM symbol after usual Phase and Amplitude Equalization

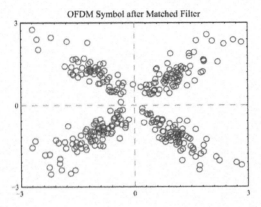

Figure 3.26: OFDM symbol after Matched Filter

a Gray mapped QPSK symbol are defined in (2.18) and (2.19), and are implemented as follow:

$$
\begin{aligned}
\Lambda_1^{[\Lambda \cdot R]} = \frac{1}{2} \max \Big\{ & \left| \Re\{\bar{y}_2\} - \frac{\hat{\rho}_{12}}{2} \right| + \left| \Im\{\bar{y}_2\} - \frac{\hat{\rho}_{12}^*}{2} \right| + \Im\{\bar{y}_1\}, \\
& \left| \Re\{\bar{y}_2\} - \frac{\hat{\rho}_{12}^*}{2} \right| + \left| \Im\{\bar{y}_2\} + \frac{\hat{\rho}_{12}}{2} \right| - \Im\{\bar{y}_1\} \Big\} \\
- \frac{1}{2} \max \Big\{ & \left| \Re\{\bar{y}_2\} + \frac{\hat{\rho}_{12}^*}{2} \right| + \left| \Im\{\bar{y}_2\} - \frac{\hat{\rho}_{12}}{2} \right| + \Im\{\bar{y}_1\}, \\
& \left| \Re\{\bar{y}_2\} + \frac{\hat{\rho}_{12}}{2} \right| + \left| \Im\{\bar{y}_2\} + \frac{\hat{\rho}_{12}^*}{2} \right| - \Im\{\bar{y}_1\} \Big\} \\
+ \, & \Re\{\bar{y}_1\}
\end{aligned}
\tag{3.21}
$$

45

$$
\begin{aligned}
\Lambda_2^{\text{IA-R}} = \frac{1}{2}\max\Big\{ &\Big|\Re\{\bar{y}_2\} - \frac{\hat{\rho}_{12}}{2}\Big| + \Big|\Im\{\bar{y}_2\} - \frac{\hat{\rho}_{12}^*}{2}\Big| + \Re\{\bar{y}_1\}, \\
&\Big|\Re\{\bar{y}_2\} + \frac{\hat{\rho}_{12}^*}{2}\Big| + \Big|\Im\{\bar{y}_2\} - \frac{\hat{\rho}_{12}}{2}\Big| - \Re\{\bar{y}_1\}\Big\} \\
- \frac{1}{2}\max\Big\{ &\Big|\Re\{\bar{y}_2\} - \frac{\hat{\rho}_{12}^*}{2}\Big| + \Big|\Im\{\bar{y}_2\} + \frac{\hat{\rho}_{12}}{2}\Big| + \Re\{\bar{y}_1\}, \\
&\Big|\Re\{\bar{y}_2\} + \frac{\hat{\rho}_{12}}{2}\Big| + \Big|\Im\{\bar{y}_2\} + \frac{\hat{\rho}_{12}^*}{2}\Big| - \Re\{\bar{y}_1\}\Big\} \\
+ \Im\{\bar{y}_1\}&
\end{aligned}
\tag{3.22}
$$

with $\hat{\rho}_{12} = \hat{H}_1^*\hat{H}_2{}^1$, $\bar{y}_1 = \hat{H}_1 y$ and $\bar{y}_2 = \hat{H}_2 y$, in opposition to (2.18) and (2.19), where the propagation channels h_1 and h_2 were supposed to be known.

Now, assuming that $\Psi = \frac{1}{2}\max\{\dots\} - \frac{1}{2}\max\{\dots\}$ is a cost function depending on the quality of the channel estimations, we can formulate (3.21) and (3.22) as:

$$
\Lambda_i^{\text{IA-R}} = \Lambda_i^{\text{II-R}} + \Psi_i(\hat{H}_1, \hat{H}_2),
\tag{3.23}
$$

where $i = 1, 2$ for a QPSK constellation.

3.6 Base Stations Channel Estimation Enhancement

According to (3.23), the performance of IA-R in comparison to II-R depends on the optimization function $\Psi(\hat{H}_1, \hat{H}_2)$, which in turn depends on the quality of the channel estimations of h_1 and h_2.

A common measure for estimation quality is the MSE (see section 3.5.1). As a first result in the investigation of the MSE, Fig.3.27 plots the MSE of $\hat{H}_1$ in an interference limited environment. The first observation, the SNR has a negligible effect on the MSE. A positive SIR means the power of the signal of interest is stronger than the interfering signal. Thus, the MSE of $\hat{H}_1$ by positive SIR should decrease and by negative SIR MSE should increase.

Fig. 3.28 shows the MSE of $\hat{H}_2$ regarding SIR. MSE of $\hat{H}_2$ decreases be negative SIR as expected.

Overlaying 3.27 and 3.28 we can get an approximation of the boundaries of the IA-R performance. Even the minimum MSE is still very high. Hence, the necessity of external support in the channel estimation process.

---

[1] $\mathcal{F}^{-1}\{H\} = h$

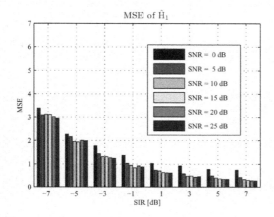

Figure 3.27: $\hat{H}_1$ vs. SIR

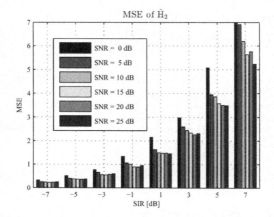

Figure 3.28: $\hat{H}_2$ vs. SIR

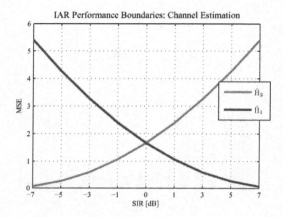

Figure 3.29: IAR Performance Boundaries

3.6.1 Serving Base Station: Holes

To assist the channel estimation process at the UE, the serving BS inserts holes in the positions of the interfering pilots, so that the quality of $\hat{H}_2$ becomes solely a noise problem (Fig. 3.30).

This method will solve the problem of the channel estimation of the interfering signal by a positive SIR, but not the problem of $\hat{H}_1$ by SIR < 0.

The serving BS must avoid a negative SIR at the OFDM symbols containing the pilots, likely by boosting the power of the transmitted signal at these positions.

Since the puncturing of N_p data symbols reduces the average power of the OFDM symbol and in some standards such as LTE-Advanced it is important to keep a constant power over all symbols, we could increase the power of the pilots of the desired signal to update the overall OFDM symbol power.

3.6.2 Interfering BS: Pilot Boosting

This approach needs a "cooperative" interfering BSs, since the distribution of the power will be adjusted in a way to permit to a user of the neighbor cell (competitor) to have

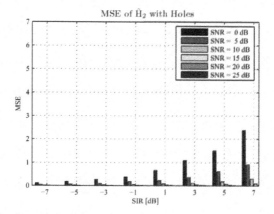

Figure 3.30: Serving BS enhancement: Holes at the pilot positions of the interfering signal

a better estimation quality.

The fact, that many papers have already discussed the adjustable power allocation between pilot and data [17, 18] and also that this approach has been already standardised in LTE-Advanced makes it very interesting.

The interfering BS will boost the power of the pilot symbols and decrease the power of the data symbols, so that the overall power at the OFDM symbols containing the pilots stay constant.

In LTE-Advanced four possible values of power offset $\delta_{power\text{-}offset}$ are predefined (Table 5.2-1 in [1]) and can be selected according to some cell-specific parameters signalled by a higher layer. In this work it is possible to give any $\delta_{power\text{-}offset}$ within its definition interval.

Lets N_p be the number of pilot symbols and N_d the number of data symbols in a given OFDM symbol, k_p and k_d the power adjusting factors for pilots and data symbols and $||s_d||^2$ and $||s_p||^2$ the power of one pilot symbol and one data symbol, respectively. Since the overall transmit power must remain constant we get:

$$N_p||k_p s_p||^2 + N_d||k_d s_d||^2 = N_p||s_p||^2 + N_d||s_d||^2, \tag{3.24}$$

since only normalized QPSK symbols for data and pilots are considered in this work we have $||s_p||^2 = ||s_d||^2 = 1$. Hence,

$$k_p^2 N_p + k_d^2 N_d = N_p + N_d = N, \tag{3.25}$$

with $N_d = 250$ and $N_p = 50$.
From (3.25) we can get $k_d(k_p)$ with $k_d^2 = (N - k_p^2.N_p)/N_d$.

Now, if k_p is the adjustable variable, we must ensure that $0 < k_p^2 < N/N_p(= 6)$

Power boosting of the interfering pilots has a double positive effect: it reduces the interference at the symbols adjacent to the interfering pilot positions, which in turn allows a better $\hat{H}_1$ and because the power of the interfering pilots is boosted, it permits a better $\hat{H}_2$.

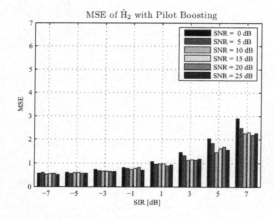

Figure 3.31: Interfering BS: Pilot Boosting

4 Simulation Results

In this section some selected simulation results are presented. The BER performance of II-R and IA-R in different interference cancellation architectures is analyzed. More precisely, we demonstrate that an IA-R without BS support in the interference cancellation process is incapable to work correctly even by best case SNR/SIR.

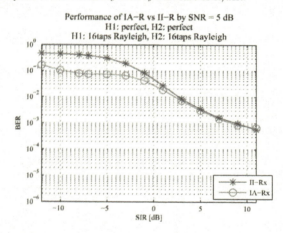

Figure 4.1: BER performance of IA-R vs. II-R by SNR = 5 dB and variable SIR and Perfect Channel Knowledge

Firstly, we start with ideal system assumption: perfect propagation channel knowledge at the receiver. As shown in Fig. 4.1, for SIR < 0 and by SNR = 5 dB, the IA-R gain in comparison to the II-R is about 10 dB. II-R starts to work correctly by a signal of interest power stronger than interfering signal power. The same BER performance for an II-R by SIR = 0 dB is already reached by an IA-R at SIR = −7 dB. By very good SIR values, the IA-R performance is at least as good as the II-R performance. In Fig. 4.2 we can see a comparison between the BER performance of IA-R and II-R by SIR values between -12 dB and 12 dB and a fixed SNR of 20 dB. Even by very good SIR the IA-R outperforms the II-R.

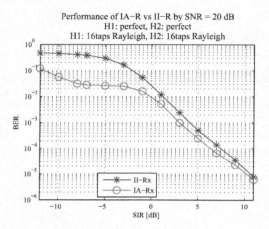

Figure 4.2: BER performance of IA-R vs. II-R by SNR = 20 dB, variable SIR and Perfect Channel Knowledge

Additionally, Fig. 4.3 illustrates, the BER performance of II-R and IA-R by different SNR and SIR values under the assumption perfect channel knowledge. IA-R always over performs II-R.

From the last observations, it becomes obvious that IA-R under perfect channel knowledge is not only a better alternative to the II-R in interference limited networks, but also in noise limited systems.

The last system assumptions stray very far from reality. Let's take a look at a more realistic model, where knowledge about the channel can be only obtained by LS estimation. As shown in Fig. 4.4 and Fig. 4.5 the IA-R in a common mobile network without BS support in the interference cancellation, specifically in the channel estimation process, is incapable to work correctly even by the best SNR/SIR values. Hence, IA-R without BS channel estimation support doesn't make any sense. Actually, IA-R and II-R are not competing against each other, since an IA-R can be any time an II-R by switching off the additive cost function in (3.23). Upon this a question arises: when does it make sense to switch on the cost function?

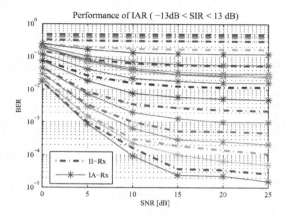

Figure 4.3: IA-R vs. II-R by Perfect Channel Knowledge

Finally, Figs. 4.6 till 4.8 depict the performance of the IA-R with BS enhancement in the interference cancellation process. As expected, both proposed solutions overperforms II-R for all possible SIR/SNR values. Actually, it is more appropriate to talk about interference cancellation architectures, since IA-R without BS enhancement is incapable to work correctly. A detailed investigation of the quantitative performance comparison between the different proposed solutions is of great interest. As a summary, the IA-R comes in use as soon as one BS, whether serving or interfering BS, guarantees a support in the channel estimation process. This observation makes IA-R much more interesting, since no need for metric for this purpose.

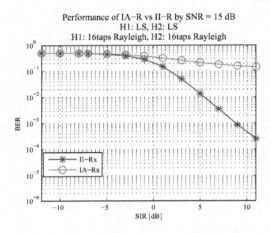

Figure 4.4: BER performance of IA-R vs. II-R by SNR = 15 dB and variable SIR and LS Channel Estimation

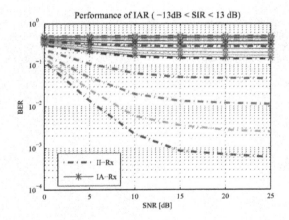

Figure 4.5: IA-R vs. II-R by LS Channel Estimation

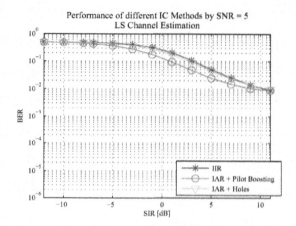

Figure 4.6: IA-R Performance with Base Station Interference Cancellation Enhancement, by SNR = 5 dB and variable SIR

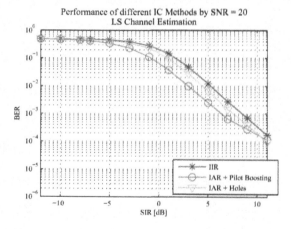

Figure 4.7: IA-R Performance with Base Station Interference Cancellation Enhancement, by SNR = 20 dB and variable SIR

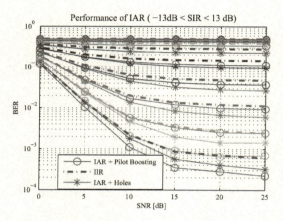

Figure 4.8: Proposed Interference Cancellation Architectures performance by variable SNR/SIR values

5 Summary and Outlook

Interference aware receiver under ideal system assumptions could be the best alternative to deal with the capacity vs. interference dilemma in heterogeneous cellular networks.

In this thesis we tested the limitations of such an intelligent user equipment under more realistic assumptions.

Starting from a basic single cell interference model we investigated the effect of different SIR on the performance of IA-R. This led to new results: IA-R performance depends on the estimation quality of each propagation channel, interferer channels as well as channel of interest.

Thus, claiming IA-R as "the solution" is far to be realistic, IA-R is part of the solution. In fact, base stations need to be involved in the interference cancellation process, specifically in the channel estimation at the UE in the DL. This requires a minimum of coordination between neighbor BSs, which is more a provider motivation problem than a scientific problem. Furthermore, the IA-R requests time synchronization among BSs and also knowledge about the modulation scheme as well as the pilot positions of the interfering signals. If all these conditions are fulfilled the IA-R can bring its promised performance gain.

Another interesting observation is: even by very weak interference signal power or rather by very high SIR, IA-R brings some performance gain in comparison to the usual interference ignorant receiver.

Combining these results and observations two solutions have been developed: serving BS enhancement or/and interfering BS enhancement in the channel estimation process at the UE in the DL.

Having these two possibilities does not only overcome the provider motivation problem but also allows the deployment of symmetric and cooperative cellular models. So, we can move from interference cancellation receiver investigation to a more global approach: interference cancellation architectures.

Future work could be directed to the inclusion of:

- symmetric hole performance

- combination with precoding and MU-MIMO
- MMSE MIMO (2 antennas)
- and use of turbo codes.

Bibliography

[1] Evolved universal terrestrial radio access (e-utra); physical layer procedures (release 10). Technical report.

[2] E. Akay and E. Ayanoglu. Low complexity decoding of bit-interleaved coded modulation for m-ary qam. In *Communications, 2004 IEEE International Conference on*, volume 2, pages 901–905 Vol.2, june 2004.

[3] Martin Bossert. *Channel Coding for Telecommunications*. Wiley, 1999.

[4] G. Caire, G. Taricco, and E. Biglieri. Bit-interleaved coded modulation. *Information Theory, IEEE Transactions on*, 44(3):927–946, may 1998.

[5] J. Hagenauer and P. Hoeher. A viterbi algorithm with soft-decision outputs and its applications. nov. 1989.

[6] Florian Kaltenberger, Rizwan Ghaffar, Raymond Knopp, Hicham Anouar, and Christian Bonnet. Design and implementation of a single-frequency mesh network using openairinterface. *EURASIP Journal on Communications and Networking*, 2010, 2010. Article ID 719523, 16 pages, doi:10.1155/2010/719523.

[7] A. Martinez, A. Guillen i Fabregas, G. Caire, and F. Willems. Bit-interleaved coded modulation revisited: A mismatched decoding perspective. *Information Theory, IEEE Transactions on*, 55(6):2756–2765, june 2009.

[8] Andre Neubauer, Jürgen Freudenberger, and Volker Kühn. *Coding theory*. John Wiley, c2007.

[9] C. Novak, P. Fertl, and G. Matz. Quantization for soft-output demodulators in bit-interleaved coded modulation systems. In *Information Theory, 2009. ISIT 2009. IEEE International Symposium on*, pages 1070–1074, 28 2009-july 3 2009.

[10] W. Rave. Quantization of log-likelihood ratios to maximize mutual information. *Signal Processing Letters, IEEE*, 16(4):283–286, april 2009.

[11] Rizwan Ghaffar and Raymond Knopp. Fractional frequency reuse and interference suppression for OFDMA networks. In *WIOPT 2010, 8th International Symposium on Modeling and Optimization in Mobile, Ad Hoc, and Wireless Networks, 31 May-4 June 2010, Avignon, France*, Avignon, FRANCE, 05 2010.

[12] Rizwan Ghaffar and Raymond Knopp. Low complexity metrics for BICM SISO and MIMO systems. In *VTC 2010-Spring, 71st IEEE Vehicular Technology Conference, 16-19 May 2010, Taipei, Taiwan,* Taipei, TAIWAN, PROVINCE OF CHINA, 05 2010.

[13] Rizwan Ghaffar and Raymond Knopp. Interference-aware receiver structure for Multi-User MIMO and LTE. *"EURASIP Journal on Wireless Communications and Networking", Volume 2011: 40,* 05 2011.

[14] Rizwan Ghaffar and Raymond Knopp. Interference suppression strategy for cell-edge users in the downlink. *"IEEE Transactions on Wireless Communications", 2011, Volume 11, N°1, ISSN:1536-1276,* 12 2011.

[15] Ron Roth. *Introduction to Coding Theory.* Cambridge University Press, New York, NY, USA, 2006.

[16] Stefania Sesia, Issam Toufik, and Matthew Baker. *LTE, The UMTS Long Term Evolution: From Theory to Practice.* Wiley Publishing, 2009.

[17] M. Simko, S. Pendl, S. Schwarz, Q. Wang, J. Colom Ikuno, and M. Rupp. Optimal pilot symbol power allocation in LTE. In *Proc. 74th IEEE Vehicular Technology Conference (VTC2011-Fall),* San Francisco, USA, September 2011.

[18] M. Simko, Q. Wang, and M. Rupp. Optimal pilot symbol power allocation under time-variant channels. *EURASIP Journal on Wireless Communications and Networking,* 225, 2012.

[19] F. Tosato and P. Bisaglia. Simplified soft-output demapper for binary interleaved cofdm with application to hiperlan/2. In *Communications, 2002. ICC 2002. IEEE International Conference on,* volume 2, pages 664–668. IEEE, 2002.

[20] Andrew J. Viterbi. An intuitive justification and a simplified implementation of the map decoder for convolutional codes. *IEEE Journal on Selected Areas in Communications,* 16(2):260–264, 1998.

[21] Sebastian Wagner. Documentation of dlsim.c. Technical report, 2012.

[22] Jun Wang, Oliver Yu Wen, Hongyang Chen, and Shaoqian Li. Power allocation between pilot and data symbols for mimo systems with mmse detection under mmse channel estimation. *EURASIP J. Wirel. Commun. Netw.,* 2011:3:1–3:9, January 2011.

[23] E. Zehavi. 8-psk trellis codes on rayleigh channel. In *Military Communications Conference, 1989. MILCOM '89. Conference Record. Bridging the Gap. Interoperability, Survivability, Security., 1989 IEEE,* pages 536–540 vol.2, oct 1989.

www.ingramcontent.com/pod-product-compliance
Lightning Source LLC
LaVergne TN
LVHW042349060326
832902LV00006B/483